AF709837

SERIE d'ECRITURE
Supplement No.3

BY ANNE-MARIE ALBIACH, in English:

Vocative Figure, trans. Anthony Barnett & Joseph Simas, Moving Letters, 1986; Second Edition: Allardyce Books, 1992
Mezza Voce, trans. Joseph Simas, Post-Apollo Press, 1988
E*tat*, trans. Keith Waldrop, Awede, 1989

Anne-Marie Albiach

A GEOMETRY

translated from the French
by Keith & Rosmarie Waldrop

BURNING DECK, PROVIDENCE

1998

SERIE d'ECRITURE is an annual of current French writing in English translation. Most issues are given to the work of a single author. The first five issues were published by SPECTACULAR DISEASES, which continues to be the sole source for European distribution and subscription. Since No. 6, the publisher has been Burning Deck in Providence, RI.
Editor: Rosmarie Waldrop

Individual copies: $10. Subscription for 2 issues: $16.
Supplements: $5
In England: £5. Subscription for 2 issues: £8. Postage 25p/copy.
Distributors:
Small Press Distribution, 1341 Seventh St., Berkeley, CA 94710
Spectacular Diseases, c/o Paul Green, 83b London Rd., Peterborough, Cambs. PE2 9BS
for US subscriptions only:
Burning Deck, 71 Elmgrove Ave., Providence RI 02906

Burning Deck is the literature program of Anyart: Contemporary Arts Center, a tax-exempt non-profit corporation.

ISSN 0269-0179
ISBN 1-886224-31-5,

Cover by Keith Waldrop

Thanks to Joseph Simas for the title "Vertical Effort in White"

CONTENTS

VERTICAL EFFORT IN WHITE

"she does not know herself, is dazzled by data"

up from the roots an added breath at work within the
strength of a measure

on the immediate verso or lengthwise, hair
swept into the night of a fate provoking mysteries in its play
across barely perceptible stage business
comes hugging knees and brushing against
hands close to innocence

*"and in the inner light they maintain
salient relations"*

prehensile and vulnerable; eyes under the sway of colors

dark doubled
in a minor key, this tactile
memory

"you there

somber"

on the median line, they seek their bodies; "you make gestures of
approach or withdrawal"   it is not in your power
to delete the outline

this opposition proves to be aleatory;   mute in
the acid air recurs so sweet their absence

filled with a fervor they woud have tried
to make tangible,

nevertheless she recalled fabrics, scents,
chains -- there where the night birds broadcast
their cries dangerous to ignore:
how they exacerbate this place;

open lips, spots, burns: such
    bodily representation was newly outlawed

"these gestures quickening; again turns away
from a tiny trace on the chest"

she trembles before their nakedness; our talk veered apart,
the uncertain day had the last word   which oscillates

the indistinct rises toward the angle   chance takes them
    under   a trajectory

"white, dazzled, she
    wastes away"

in this place where images fade a
plot of land is its own proof;   a
reminiscence embraced as if we were
waiting, exhausted, for sunrise

she abstracts the object; she deflects the gesture that
catches in the gap between, faced with food and its names in
the extreme; a greed for offerings, words of sacrifice

"thus proffered, the telluric image, a disparity"
"she engendered her son in stubborn silence or generating
a stylistic figure"   and a memory likely to come undone at
the slightest change of vowel, an abstraction of origins; her
face reflected the homonym of an earlier loss   the cues
this archaic episode carries consequence

the law of succession the blankness of symbols
vertebral distance   turned
pale and broke this kind of logic

INCANTATION

"I ate from the hand of a God in order to come forth and continue on this fractured earth"

She impregnates my face. Her hair taken, and in the veins addicted blood, coming from elsewhere, alternate liquid element.

Insidious, she disowns a journey, graphic signs, eyes, in this meticulous account which no longer frightens them at all.

"memory borrows from flowers and forest such exact art"

On every side strokes appear, whence a voice approaching incantation.

Billow, obscure destitution, she withdraws into silence, in the least degree a stranger.

In their childhood, blue cuffs;

"milk of generations"

immobility is constrained

it thwarts an otherness and the other's glance connects with this elaboration.

over the years, a perverse goal's elements.

Such investigation strips bare an indeterminate time, abasing the relapsed, gestures from this time forward.

Dark shadow lets a body fall, recurring drop into opacity:

"the cold is stamped with the sleeper who, roused, restores me to life"

Three outlines and a pallid erudition. She gives birth in the lineage of chance; premonition of data: night annihilates objects of an incantatory solitude, thinned by sleep

"this excitement of the first days"

An illicit body, nudity in the breath. *They*
hereafter in rumor.

"at the sight of blood he swooned"

They will come no more. In dorsal
hues, they seek each other at break of day.

I dressed this unprecedented wound in its last stage. The night was gasping and its fruits even to oblivion. A sketch on the bosom, this color cast anew on the earth: heat suddenly in the margins. Repetition of absences

"this complicity
to the point of injury"

FIGURES OF MEMORY

the term specifies the movement

flagrant
parallel begetting of a form

the internal figures dissolve

gestures of destruction unstable

"at first stroke stifled"

previous words

at that moment

and she turns away : he stands still

she had seen them daily
in the reiteration of anonymity

laws --

obedient to the pressure within
"struck dumb by scandal"

they spring up

where their sight grew dim

"she dared not infringe that aversion"

confined at the boundary

outside implied a prohibition

the notch *scans*
the same alliteration

would they thereafter have enumerated the ensemble

o

"she talks of coming seasons"
ousting alterity

the reply in trembling
a punctuation

o

violence
no end
or the play of outmoded numbers

by sheer force of memory

and this time he went away

in his limbs' gravity

"hurt breast"

from the liquid element the myth sprang up
"mouth open beyond the bruises"

subjected to this elucidation

their word

alternative

"sleep
where the censor menaces"

ablution

this chill in the corridors

rough draft

archaic
failure of eloquence

"hand reaching for forbidden fruit"

signs
on the bias tongues laminated nightly
such theatricality of indictment

" *dorsal exacerbation*"
of day's double gash

"a geometry"

NOTES

Anne-Marie Albiach is the author of *Flammigère* (Siècle à mains, 1967), E*tat* (Mercure de France, 1971), *Mezza Voce* (Flammarion, 1984), *Anawratha* (Spectres Familiers, 1984), *Figure Vocative* (Lettres de Casse, 1985; Fourbis, 1991), and*Travail vertical et blanc* (Spectres Familiers, 1989). With ClaudeRoyet-Journoud and Michel Couturier, she co-edited the magazine *Siècle à mains*, where she first published her legendary translation of Zukofsky's "A-9."
Interviews can be found in *Action Poétique* #74 (1978), *Arpentage de la poéisie contemporaine* by Jacques Darras (in'hui, 1987) and, in English, in *Acts* #4 (1985). Jean-Marie Gleize has published a book on her work: *Le Théatre du poème: vers Anne-Marie Albiach* (Paris: Belin, 1995).

Keith Waldrop's most recent books of poems are *Analogies of Escape* (Burning Deck), *The Silhouette of the Bridge* and *The Locality Principle* (both Avec Press). Sun & Moon published his novel, *Light While There Is Light* (1993). He has also translated Claude Royet-Journoud, Dominique Fourcade, and Jean Grosjean.

Rosmarie Waldrop's most recent books are *Another Language: Selected Poems* (Talisman House) and *A Key Into the Language of America* from New Directions, which will also publish *Reluctant Gravities*. Station Hill has published her novels, *The Hanky of Pippin's Daughter* and *A Form/ of Taking/ It All*. Translations include Edmond Jabès, Jacques Roubaud, Friederike Mayröcker, and Elke Erb .

SERIE D'ECRITURE

No. 1: Alain Veinstein, Archeology of the Mother (tr. Tod Kabza, Rosmarie Waldrop), 1986

No. 2: Emmanuel Hocquard, Late Additions (tr. Connell McGrath, Rosmarie Waldrop), 1988

No. 3: Anne-Marie Albiach, Marcel Cohen, Jean Daive, Dominique Fourcade, Jean Frémon, Paol Keineg, Jacqueline Risset, Jacques Roubaud, Claude Royet-Journoud (tr. Anthony Barnett, Charles Bernstein, Lydia Davis, Serge Gavronsky, Rachel Stella, Keith Waldrop, Rosmarie Waldrop), 1989

No. 4: Anne-Marie Albiach, Olivier Cadiot, Danielle Collobert, Edith Dahan, Serge Fauchereau, Dominique Fourcade, Liliane Giraudon, Joseph Guglielmi, Vera Linhartova, Anne Portugal (tr. Charles Bernstein, Norma Cole, Robert Kocik, Natasha, Ron Padgett, Keith Waldrop, Rosmarie Waldrop), 1990

No. 5: Joseph Guglielmi, Dawn (tr. Rosmarie Waldrop), 1991

No. 6: Jean Daive, A Lesson in Music (tr. Julie Kalendek), 1992

No. 7: Pierre Alferi, Jean-Pierre Boyer, Olivier Cadiot, Dominique Fourcade, Jean Frémon, Jean-Marie Gleize, Dominique Grandmont, Emmanuel Hocquard, Isabelle Hovald, Anne Portugal, Jacques Roubaud, James Sacré, Anne Talvaz, Esther Tellerman (tr. David Ball, Norma Cole, Stacy Doris, Paul Green, Tom Mandel, Pam Rehm, Cole Swensen, Keith Waldrop, Rosmarie Waldrop), 1993

No. 8: Paol Keineg, Boudica (tr. Keith Waldrop), 1994

No. 9: Marcel Cohen, The Peacock Emperor Moth (tr. Cid Cor-man), 1995

No. 10: Jacqueline Risset, The Translation Begins (tr. Jennifer Moxley), 1996

No. 11: Alain Veinstein, Even a Child (tr. Robert Kocik & R. Waldrop)

Supplements:

No. 1: Claude Royet-Journoud, i.e. (tr. Keith Waldrop), 1995

No. 2: Pascal Quignard, Sarx (tr. Keith Waldrop), 1997

No. 3: Anne-Marie Albiach, A Geometry (tr. K. & R. Waldrop), 1998